When I grow up I want to become a petroleum engineer

Written By John O'Connell

Illustrations By Sarah Kapp

I would like to dedicate this book to my beautiful children Stephanie, Jill, Callum and little Tara.

And of course to my father, the engineer, who inspired me to write this children's book.
Thank you also to my wife Ludmila for being so patient with me.
Finally, to my mother who gave me the strength of a lion!

Hi you guys! My name is Johnny! What's your name? I'm 8 years old. I was born in Calgary Canada. Where were you born? My Dad is a "Petroleum Engineer" but I don't really know what that means. What does your Dad do? My Dad always told me that I should work hard, whatever my job is, and take pride in my work! I didn't know what the word "pride" means so my Dad told me that it means when you are proud of something. I am proud of my Mom and Dad. I am also proud of my older sister who is really good at basketball! What are you proud of?

My mom's name is Ann and my dad's name is Patrick and that seems funny to me becuase I only call them mom and dad!! HaHaha! My dad is from Ireland and my mom is from England. My last name is O'Connell. What are your parent's first names? What is your last name? Where are your parents from?

The word "petroleum" comes from two Latin words. Latin is an old language that is not spoken in the world anymore but many of the words in the English language came from this language. "Petra" in Latin means "rock" and "oleum" means oil! So if you put them together they mean "rock oil"! This makes sense because we get petroleum products from the ground and I will explain the whole process to you.

Liquid **Solid** **Gas**

The first thing you need to know is that the petroleum industry is about "oil" and "gas". This oil and gas comes from the ground below us. Sometimes it can be found near the surface but more often it is found far below the surface! Oil is a "liquid" and gas is what we call "vapor" like the air we breathe or when water turns into steam when you boil water on the stove. I know you can't see air but it is really all around us and we breathe it in and out all day and night! (Please look at the photos and discuss them with your mum and dad!)

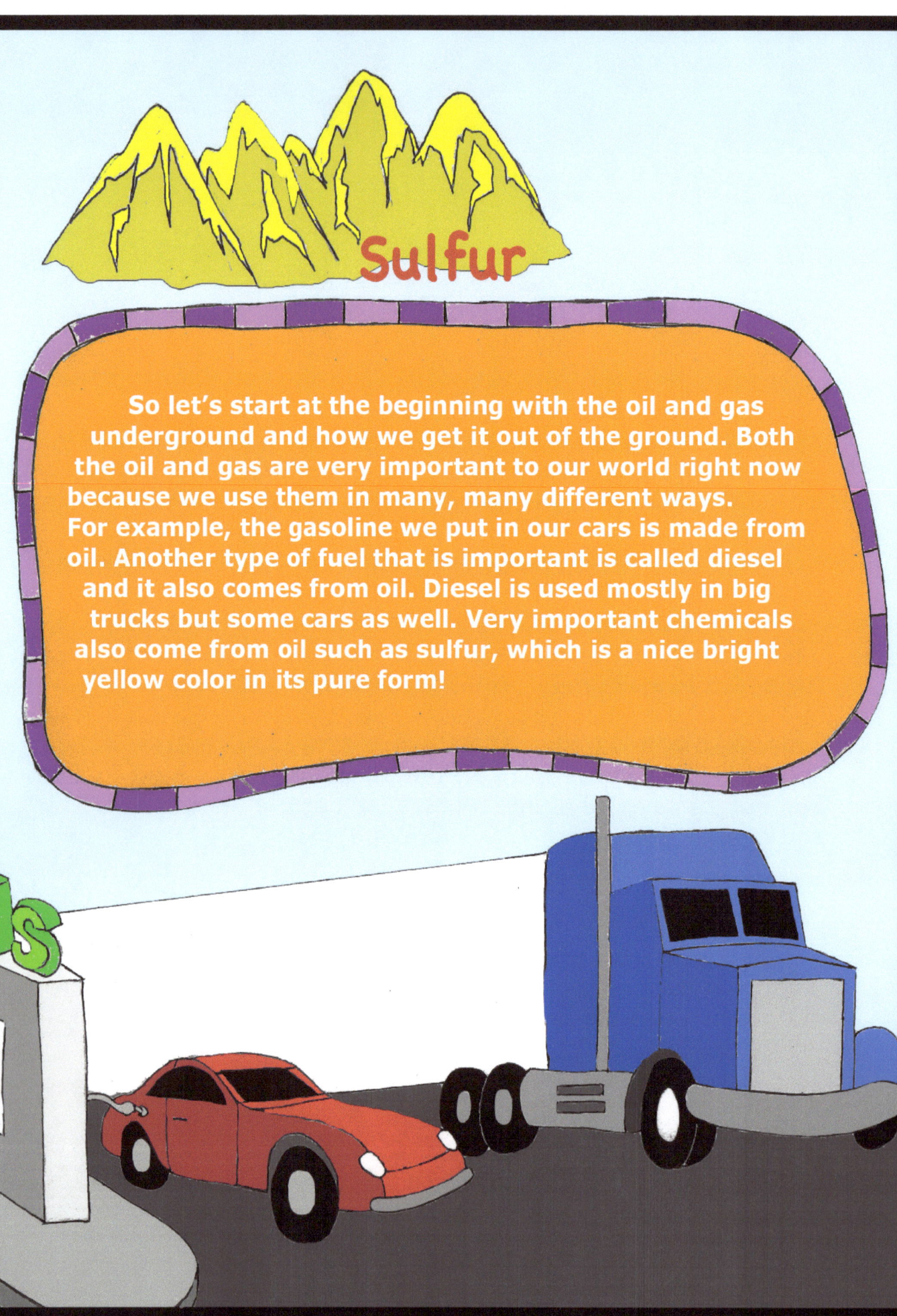

So let's start at the beginning with the oil and gas underground and how we get it out of the ground. Both the oil and gas are very important to our world right now because we use them in many, many different ways. For example, the gasoline we put in our cars is made from oil. Another type of fuel that is important is called diesel and it also comes from oil. Diesel is used mostly in big trucks but some cars as well. Very important chemicals also come from oil such as sulfur, which is a nice bright yellow color in its pure form!

Gas is also very important and is what many people use to keep our houses warm in the winter. We burn the gas in our furnace to warm our houses and buildings. A fan helps move the warm air around the house. Sometimes, the gas is burned to heat water which is then used to heat the hosue. No we don't spray warm water all over the house! Hahaha! We use a radiator to transfer the heat from the hot water to the house! The burned gas and air go out through a chimney because these gases are very dangerous to our health and we must keep it away from the air we breathe!!

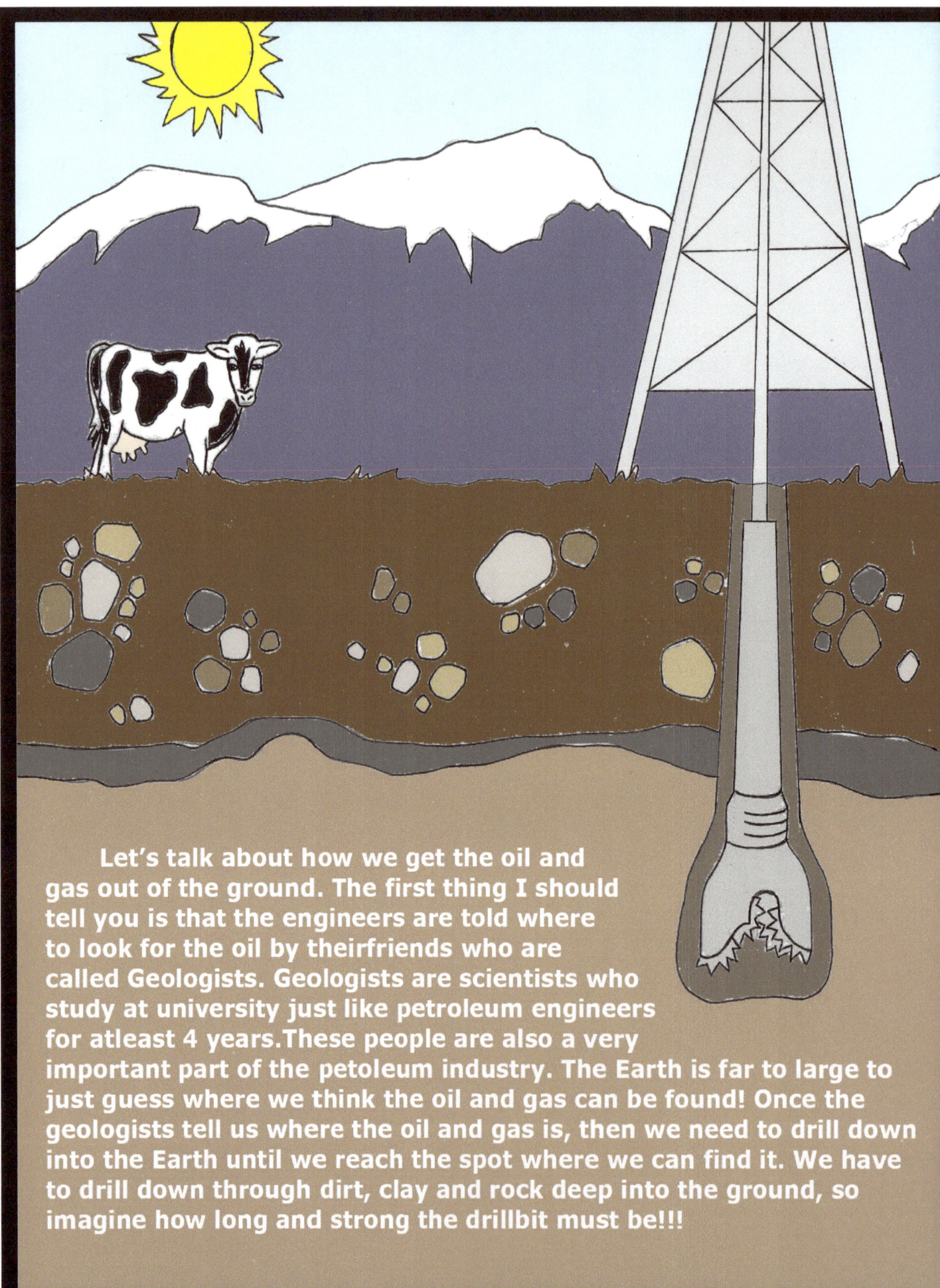

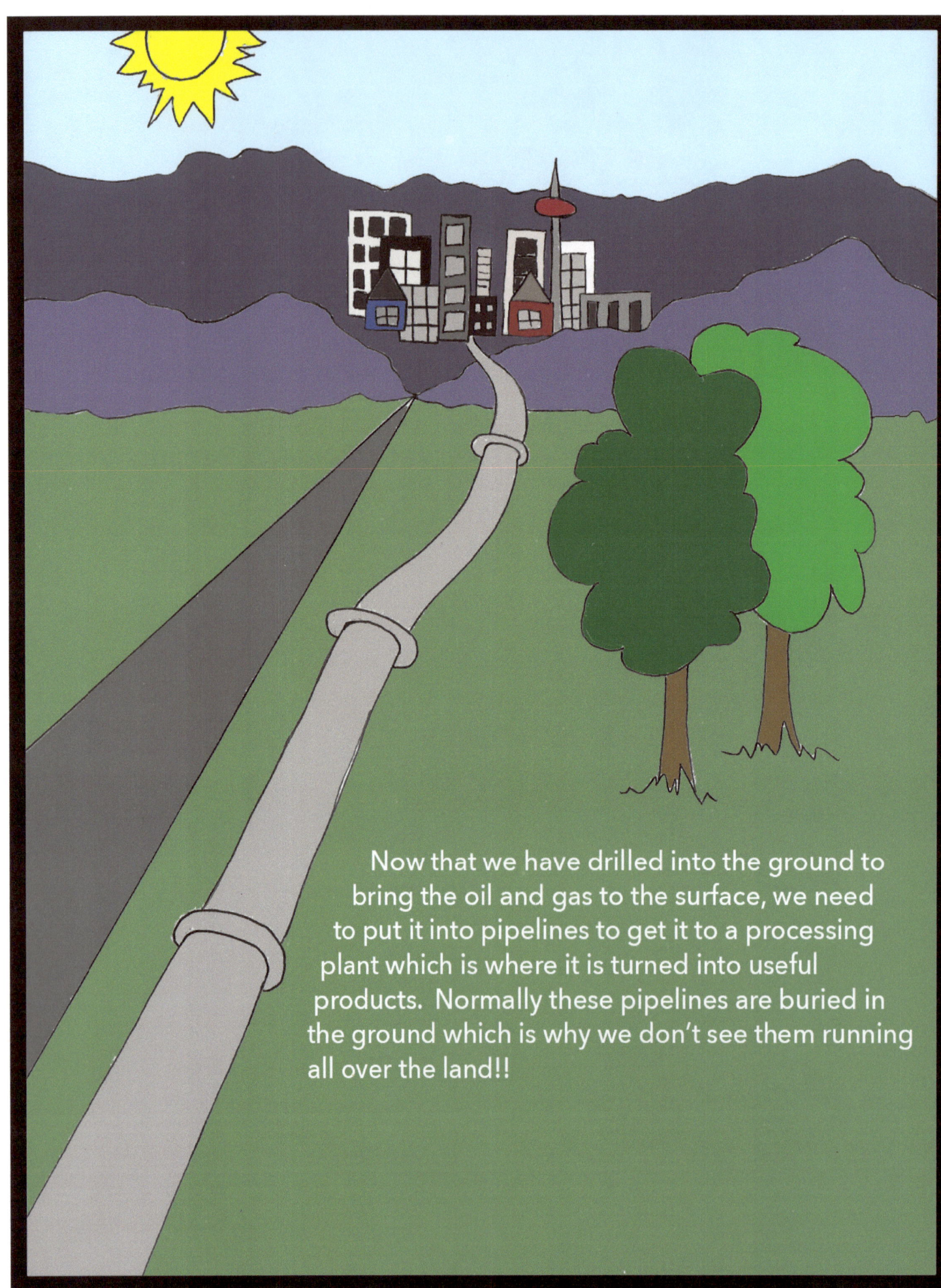

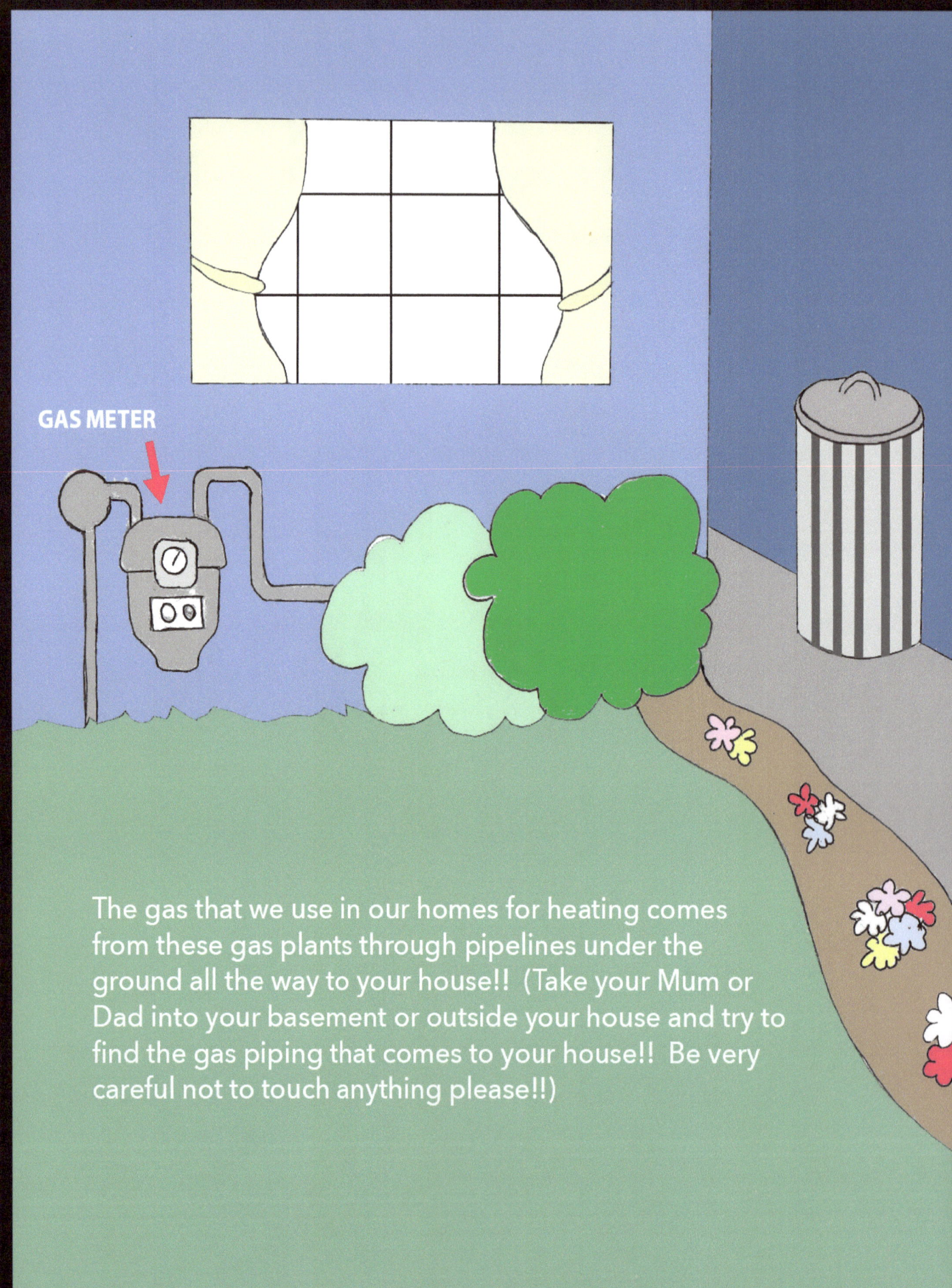

The gasoline that is made at the refinery is brought to the gas stations by large trucks. These trucks put the gasoline into large underground tanks at the gas stations. Have you ever wondered where the gas comes from when your Mum or Dad is putting gasoline in your car?? Well, now you know for sure!!! It is not really so complicated but it is an interesting journey for the oil in the ground to make it into your car!

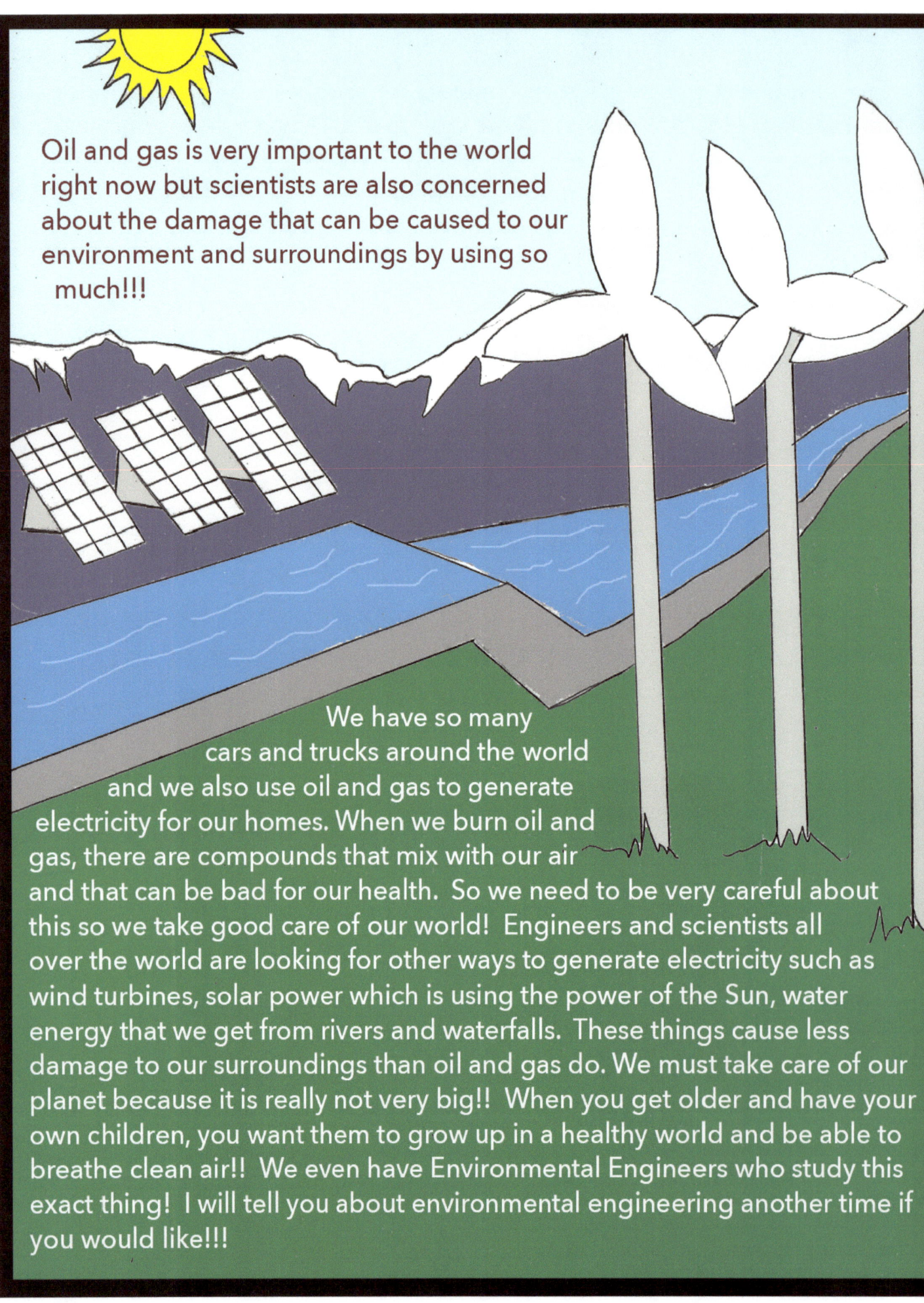

Oil and gas is very important to the world right now but scientists are also concerned about the damage that can be caused to our environment and surroundings by using so much!!!

We have so many cars and trucks around the world and we also use oil and gas to generate electricity for our homes. When we burn oil and gas, there are compounds that mix with our air and that can be bad for our health. So we need to be very careful about this so we take good care of our world! Engineers and scientists all over the world are looking for other ways to generate electricity such as wind turbines, solar power which is using the power of the Sun, water energy that we get from rivers and waterfalls. These things cause less damage to our surroundings than oil and gas do. We must take care of our planet because it is really not very big!! When you get older and have your own children, you want them to grow up in a healthy world and be able to breathe clean air!! We even have Environmental Engineers who study this exact thing! I will tell you about environmental engineering another time if you would like!!!

Photo 1: Sulfur Block

Photo 2: Pipeline

Photo 3: Gas Plant (Alberta, Canada)

Photo 4: Gas Plant (Astrakhan, Russia)

Photo 5: Catastrophic Failure of Stack

Photo 6: Pollution.
"Petroleum engineers must work hard to minimize pollution to the environment"

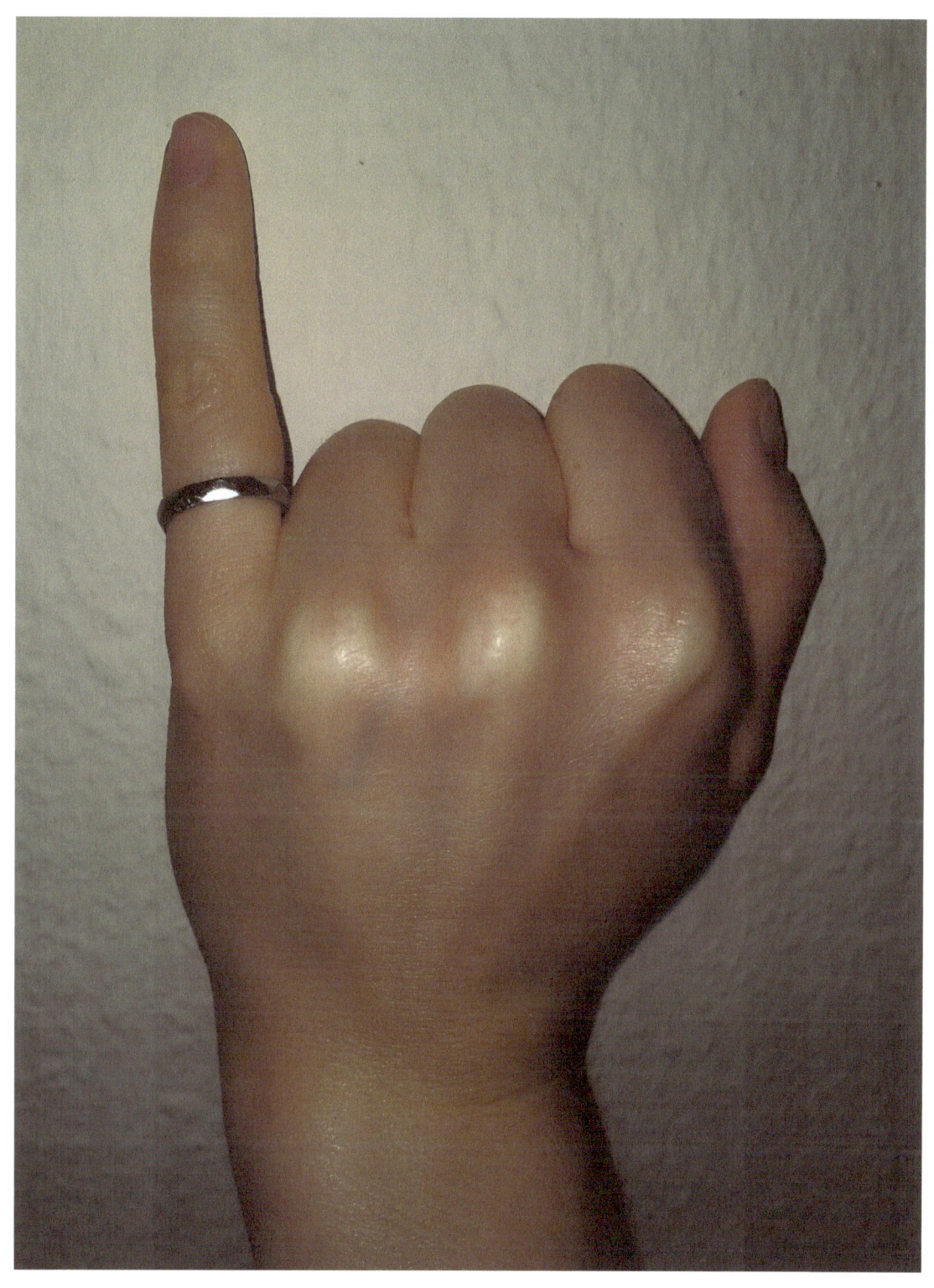

Photo 7: The "Iron Ring" worn by Canadian engineers

About The Author

John O'Connell is a chemical engineer and a practising member of the Association of Professional Engineers and Geoscientists of Alberta (APEGA). John has been working in the petroleum industry for 30 years and has worked as an operator, process engineer and engineering consultant. John also holds a Third Class Power Engineering Certificate from the Alberta Boilers Safety Association (ABSA). John graduated from the University of Calgary in 1988 and has spent the bulk of his career in the field of sulfur recovery. John was the founder of Sulfur Recovery Engineering in 1998 and since that time, the company has been responsible for decreasing global sulfur dioxide emissions by over 1,000,000 tonnes! Since 2012, John has been the lead sulfur specialist engineer for Saudi Aramco and has assisted in designing environmentally responsible sulfur recovery units that are capable of recovery efficiencies of 99.9+ percent. John also has a patented process for sulfur recovery that can achieve recovery efficiencies of 99.99+ percent.

www.ingramcontent.com/pod-product-compliance
Lightning Source LLC
Chambersburg PA
CBHW051828210526
45473CB00005B/1788